NATIONAL AUDUBON SOCIETY
POCKET GUIDE

A Chanticleer Press Edition

Ann H. Whitman
Editor

Kenn Kaufman
John Farrand, Jr.
Consultants

Eastern Region

FAMILIAR BIRDS OF NORTH AMERICA

Alfred A. Knopf, New York

This is a Borzoi Book
Published by Alfred A. Knopf, Inc.

Prepared and produced by Chanticleer Press, Inc.,
New York.
Color reproductions by Nievergelt Repro AG, Zurich,
Switzerland.
Typeset by Dix Type, Inc., Syracuse, New York.
Printed and bound by Toppan Printing Co., Ltd.,
Hong Kong.

Published October 1986
Ninth printing, November 1995

Library of Congress Catalog Card Number: 86-045588
ISBN: 0-394-74839-5

Contents

How To Use This Guide

Of all wild creatures, birds are the most conspicuous and colorful. Even in our crowded cities, there are many kinds of birds, easily noticed and easily identified. Learning to identify birds is an enjoyable pursuit—an end in itself or the first step in an intensely absorbing hobby that can last a lifetime.

Coverage

This new guide covers 80 of the most frequently encountered and abundant birds of the East. Our geographical range is bounded by the Atlantic Ocean on the east, and on the west by 100th meridian, extending from Edwards Plateau in Texas northward through Oklahoma and westward through eastern Colorado, Wyoming, and Montana, and Canada, along the eastern foothills of the Rocky Mountains. (Thus, the boundary roughly follows the Rockies.) The companion volume to western birds covers species west of this boundary.

Organization

This easy-to-use guide is divided into three parts: introductory essays; illustrated accounts of the birds; and appendices.

Introduction

As a basic introduction, the essay "Identifying Birds" outlines the field marks you should notice when you look at a bird. "Bird-watching" offers expert advice on when, where, and how to look for birds. Finally, "Attracting Birds" provides tips on how to make your backyard a haven for colorful songsters.

The Birds — This section contains 80 color plates arranged visually by the color, shape, and overall appearance of each species. Here are a selection of common water birds, shorebirds, birds of prey, game birds, land birds, and songbirds. Facing each illustration is a description of the species' important field marks as well as information about its voice, habitat, and range. A map showing nesting and winter ranges supplements the range statement. The introductory paragraph provides information about behavior, nesting habits, species, and close relatives.

Appendices — Featured is an essay on the 35 families of birds represented in this book (out of a total of 69 North American families). Knowing family traits helps to identify quickly birds covered in this guide as well as other close relatives. The last feature is an illustrated glossary defining terms that may be unfamiliar.

Whether your home is in a remote countryside or in a populous city, you will derive hours of pleasure observing and learning about birds.

7

Identifying Birds

When you see a new and unfamiliar bird, what points should you notice to help you in naming it? By going from important general features to specific ones, you will be able to narrow the identification down to a few clear choices.

Size   Exact size in inches is very difficult to judge in the field. You can usually get a general impression, however, if you compare the size of a new bird to that of something familiar, such as a House Sparrow, a robin, a crow, or some larger bird.

Shape   Experts can name most birds by shape alone. Even for beginners, learning to analyze shape helps to place a bird in the correct group.

Overall, is the bird slim or chunky? Are its neck and legs long or short? Is its tail short, medium, or long? Is it forked at the tip, square-ended, rounded, or pointed? Bill shape is one of the most important features in placing a bird in the correct family: Is the bill short or long? Thick or thin? Straight or curved? When the bird flies, are its wings long and pointed, short and rounded, or some other shape? What is the shape of its head? Does it have a crest?

8

Behavior	What a bird does often helps reveal what it is. Try to notice the way a bird perches, how it forages for food, what it eats, how it flies, and so on.
Color and Pattern	Try to look at the bird systematically, noting the crown, face, throat, underparts, wings, tail, and upperparts. Is the crown broadly striped, finely streaked, or plain? Is there a central spot of color, a pale ring around the eye, a stripe over the eye, or a dark ear patch? Does the throat contrast with the face, or with the breast? Is it framed by dark "mustache" stripes? Are the underparts streaked (lengthwise), barred (crosswise), spotted, or plain? Is there any contrasting color under the base of the tail? Are there contrasting wing bars? How many, and what color? Are there contrasting outer tail feathers, a band at the tip, or spots at the corners? Is the back streaked or plain? Is there a pale patch on the rump (just above the tail)?
Voice	In some difficult groups, voice is the best way to separate similar species. And for most birds, it is useful to make a notation of any songs or calls that you hear.

Bird-watching   It is possible to see birds practically anytime and anywhere, but certain times of day, places, and seasons are undeniably better. Knowing when to look and how to search can make your bird-watching more exciting. There is no need to travel far; wherever you live, there are undoubtedly some good birding areas nearby. Local birders may direct you to favorite spots, but even without such advice you will soon recognize good places.

Habitat   All birds need food, water, and shelter, but no two species have precisely the same needs. Thus variety in the habitat is the key to finding a variety of birds. For example, a mixed forest is usually better than a stand of just one kind of tree; a forest with trees of all ages, brushy undergrowth, clearings, and dead timber will be even better. The edge of a forest—or any place where one habitat meets another—is often a spot where birds are most numerous and easiest to see.

Water improves a good birding area, not only for aquatic species but also for land birds, which may come to drink. Here again, variety is important: Ponds or rivers with heavy vegetation, marshy inlets, and muddy estuaries are always promising.

Birds also gather where food is locally abundant, and

concentrations of fruiting or flowering plants may attract many different species. Garbage dumps are often excellent places to study gulls and other scavengers. A few species are adapted to simple environments, such as short-grass fields or the open ocean. Visit as many habitats as possible, concentrating on those that are richest in variety but not neglecting the others.

Time When you go birding, consider the time of day. Most land birds are active very early in the morning and again in the evening, but they can be very quiet at midday. In cold or wet weather, small birds tend to remain active all day. Water birds have schedules that may be governed more by the tides than by time. Many rest at high tide and feed when the water rises or falls.

Seasons Each season of the year has its own advantages for bird-watching. Winter can be a good time to begin, because—especially in the north—fewer species are present, allowing one to learn the birds gradually. Wintering birds tend to form flocks at good feeding areas, such as hedgerows and orchards. Spring is the favorite season for many bird-watchers. In the East, most spring migration occurs between early April and late May; the northbound migrants are in their most colorful plumage, and many sing on their way. Summer is the breeding season for most species, and you can watch birds build their nests and raise their young. The fall migration lasts much longer than that of spring; the first southbound shorebirds appear by early July, and the last of the waterfowl are still moving south in early December. There are more birds then, because populations are swelled by the young hatched in summer. Autumn is challenging, because many species are harder to identify in fall plumage.

Birding Techniques Try to walk slowly and smoothly, because birds are likely to be alarmed by sudden motions. Learn to watch vegetation for movements that reveal the feeding actions of small birds.

Using Binoculars The best way to see field marks is to use binoculars, but many beginners have trouble with them. A good technique is to stare fixedly at a bird while raising the binoculars into your line of sight; this will aim them correctly. Mastering this method will make your bird-watching more rewarding.

In choosing binoculars, look at the specifications—for example, 7×35. The first number is the magnification; in this example, the image will be enlarged seven times.

The second number is the diameter of the objective (forward) lenses, which governs the amount of light allowed in. Higher-powered binoculars need more light, and generally this second number should be at least five times the first. This rule is less important in very high-quality (and high-priced) binoculars. You should also consider the minimum focusing distance: To look at a small bird just a few yards away, you will need to have binoculars that allow you to focus that close.

Calls and Songs Birds utter two basic kinds of vocalizations—songs and calls. Songs, usually complex, are used mainly by adult males during the breeding season to establish territories or attract mates. Calls are usually simple notes, single or repeated, given at all seasons to express alarm or to maintain contact. All songs and most calls are distinctive; concentration and practice are the keys to recognizing them. You can often fix a bird's voice in your memory by describing it to yourself, using the transcriptions in this book as a guide. Practice calling birds to you: Squeaking or "shushing" noises often arouse curiosity. Above all, listen. Experienced observers find and identify many of their birds by sound.

Attracting Birds

Knowing what birds need is helpful when you go looking for them, and you can apply this knowledge to bring birds to your garden. By providing food, water, and shelter, you can easily attract a lively assortment of birds year after year.

Feeders   Bird-feeders range from the simple to the elaborate. One common arrangement is an open seed tray with tiny drain holes in the bottom, and perhaps a small roof to keep off rain and snow. Seed feeders attract thick-billed birds like finches and sparrows, as well as chickadees, nuthatches, and others. Provide both small and large seeds; thistle, millet, and sunflower seeds make an ideal combination, or you can add cracked corn or raisins for variety. A piece of fruit on a spike beside the feeder may bring in robins and mockingbirds.

Some people like to put out suet, which should be enclosed in wire mesh to keep large birds or squirrels from making off with the whole piece. Suet feeders attract tree-climbers, like woodpeckers and nuthatches; many birds find peanut butter an acceptable substitute. Hummingbird feeders can liven up any garden in summer, and may be used all year in warmer regions. These feeders—red tubes of plastic or glass, filled with

14

sugar-water—are sold at garden-supply stores. Regular maintenance and cleaning are important.

Squirrels   At feeders, squirrels can be pests, but they may often be foiled by strategic planning. Place the seed tray atop a smooth pole and attach a broad cone or disk of sheet metal to the pole some distance below it. Some hanging feeders are also safe.

Baths   Birds like to visit a garden bath. A bird-bath should be very shallow with a sloping bottom, and should be cleaned regularly. Resident birds will quickly find a bird-bath, but to attract transient birds it helps to rig up a small fountain so that the sound of water is audible.

Shelter   Ultimately, the best way to bring birds to your yard is to plant the trees and bushes that provide natural shelter. Evergreens, thick bushes, and vines provide a year-round haven; mulberry, multiflora rose, and other plants that also furnish berries are doubly attractive to many species. Put your feeder near enough to these plants that the birds can easily flee from danger, but not so close that cats or other predators can ambush unwary birds. And if you allow a quiet corner to grow up in annual weeds, it will be perennially popular with the birds.

THE BIRDS

Herring Gull *Larus argentatus*

The abundant and familiar Herring Gull is at home in almost any setting, from the wilds of ocean coasts to sluggish streamsides and crowded cities. These energetic scavengers are often seen whirling and turning in the air off the sterns of fishing boats and ferries, or wheeling hopefully above garbage dumps, looking for a meal.

Identification 23–26″. Large with white head and neck, gray wings and mantle, and extensive black wing tips with white spots. Bill large and yellow, with red dot near tip on mandible (lower half). Winter birds have dusky streaking on head and neck. Juveniles and immatures (first- and second-year birds) grayish-brown below and scaly gray above.

Voice A loud, trumpeting *keeyow, kyow-kyow-kyow.*

Habitat Ocean shores, bays, rivers, lakes, ponds, and wharves; also seen in plowed fields, cities, and landfill sites.

Range Alaska to Newfoundland, south to British Columbia, Montana, Great Lakes region, and New England; along Atlantic Coast to Florida and Gulf; Mississippi River; absent from mountains and much of Great Plains.

18

Laughing Gull *Larus atricilla*

The Laughing Gull is the common summer "sea gull" along much of the southern Atlantic Coast and Gulf shores. Like its larger relative, the Herring Gull, this species is a practiced beggar with an eye for seaside picnics. In recent years, the Laughing Gull's numbers have declined, perhaps as a result of the recreational development of salt-marsh areas and other suitable nesting sites.

Identification 16–17". Medium-size gull with dark gray mantle and black hood in summer; hind edge of wing white, and wing tips solid black; legs and bill dark red (bright red in some birds). Winter adults lack black hood. Juveniles dark brown with white rump.

Voice A loud, high *ha-ha-ha-ha-haah-haah-haah*.

Habitat Coastal areas, especially estuaries, bays, and salt marshes; rare inland.

Range Breeds along coast from Maine to Florida and Gulf of Mexico; winters from Virginia south, occasionally farther north.

Common Tern *Sterna hirundo*

These common birds are adroit in flight and skillful divers, plummeting from a height of many feet to snatch shrimp and small fish from the water. Terns nest in large, noisy colonies on protected beaches, and they heatedly defend their territories from all comers, including people. Common Terns are very sensitive to disturbance, and human disruption of nesting colonies has caused a large decline in their numbers.

Identification 13–16″. Adult in breeding plumage white with black cap and pale gray mantle; often washed with pale gray below; tail deeply forked, with black outer edges. Bill red with black tip; legs red. Winter adults lack most of black cap. Juveniles mottled, quickly becoming gray-backed.

Voice A long *keee-yar* and a harsh, stacatto *kip-kip-kip*.

Habitat Coastal areas, lakes, ponds, and rivers.

Range Breeds from Alberta east through Great Lakes to Labrador and New England; south along Atlantic Coast to Virginia. Winters mainly from Florida to South America; rare in winter in southernmost states.

22

Killdeer *Charadrius vociferus*

The adaptable, noisy, and conspicuous Killdeer seems to love to call attention to itself. In nesting areas, when challenged by the presence of a predator, this bird performs its famous distraction display: Feigning injury, it drags itself unsteadily across the ground, dragging a "broken" wing behind it. This behavior leads the predator away from nestlings or eggs; once the intruder is at a safe distance from the nest, the malingerer takes flight, uttering its shrill cry.

Identification 9–11". Robin-size. Grayish-brown above, white below, with 2 distinct black breast bands; rufous rump and tail conspicuous in flight. Bill thin, black.

Voice A shrill, ringing *kill-deeeer, kill-deeer,* repeated incessantly; also a shorter *dee, dee, dee.*

Habitat Fields, pastures, mud flats, beaches, and other open areas near water.

Range Breeds from central Alaska to Newfoundland, south to Mexico. Winters inland in mild areas; populations from New Jersey and Ohio southward nonmigratory.

24

Common Snipe *Gallinago gallinago*

This shy and secretive marsh denizen has achieved a certain fame for its "winnowing" courtship display; on breeding grounds the bird performs a spectacular dive, its outer tail feathers producing an eerie, hollow, dull whistling sound. When flushed, the Common Snipe takes off in an abrupt, fast zigzag flight. These birds migrate in flocks at night, but by day the Common Snipe is usually a loner.

Identification
10½–11½". Long-billed, slender brownish shorebird; mottled upperparts and head have prominent buffy stripes; belly white, breast and flanks with heavy bars and spots. In flight, has long, pointed, dark wings.

Voice
A sharp, grating *scaip* when flushed; a high *wheet-wheeet* on breeding grounds.

Habitat
Marshes, ponds, wet fields, and peat bogs; rarely in salt marshes.

Range
Breeds from Alaska to Labrador, south to California, Indiana, and Massachusetts; sometimes farther south. Winters from southern part of breeding range south.

Spotted Sandpiper *Actitis macularia*

This well-known American shorebird is found nearly throughout the continent. When feeding, it constantly bobs its tail up and down, earning it the nickname "teeter-tail." This conspicuous habit and its distinctive style of flight, in which brief glides alternate with rapid bursts of shallow wingbeats, make this sandpiper easy to recognize.

Identification | 7–8″. Robin-size. Trim; gray-brown or olive-brown above; breeding adults have heavy black spotting below. White wing stripe conspicuous in flight. Winter birds and juveniles white with brown smudge on neck and breast.

Voice | A clear *peet-weet* or *weet-weet;* also gives a soft trill.

Habitat | Anywhere near water, from beaches to open prairie potholes and wooded areas with ponds.

Range | Breeds from N. Alaska east to Cape Breton Island, south to California, N. Texas, Kentucky, and Virginia. Winters from S. United States to South America.

Sanderling *Calidris alba*

These plump little sandpipers are gregarious; they often form pure flocks of several dozen birds, but they usually feed along the shore in hurried clusters of three to six. Sanderlings are delightful to watch as they chase up and down a beach on stiff legs to probe for tiny mollusks and crustaceans washed up by ocean waves. These birds are more tame and approachable than most sandpipers.

Identification 7–8½". Small sandpiper. Breeding adults have rufous head, back, wings, and upper breast, with blackish mottling; belly white. Winter plumage pale gray above, white below. Bill and legs black. Bold white wing stripe conspicuous in flight.

Voice Flight call a sharp, high-pitched *kip* or *kit*.

Habitat Primarily along sandy beaches; also on tidal flats, lakeshores, and sandbars.

Range Breeds in Far North in tundra of Canada and Greenland along Arctic Ocean and northern Hudson Bay; winters along both coasts and south to South America.

American Coot *Fulica americana*

This aquatic species appears ungainly as it takes off from the water, pattering across the surface on its odd-looking lobed feet to gain enough momentum for flight. A skillful diver and swimmer, the American Coot feeds on aquatic vegetation, but can be induced to come out of the water for a handout of grain or scraps of bread.

Identification 13–16″. Adult slate-gray with black head and neck, setting off pale, ivory-colored bill and frontal shield (shield has red spot, conspicuous at close range). Immature similar to adult but paler, with dull bill.

Voice A variety of cackles, clucks, and low croaking notes.

Habitat Open freshwater marshes, ponds, and lakes; also in salt water in winter, when freshwater habitats are frozen.

Range Breeds from British Columbia and N. Alberta to S. Ontario and New England, south to Mexico and Florida. Winters from Illinois and Massachusetts to Florida and Gulf Coast; also along Pacific Coast and in the Southwest.

Pied-billed Grebe *Podilymbus podiceps*

This species is the most common member of its family in North America. When alarmed, the Pied-billed Grebe quietly makes itself inconspicuous; it sinks slowly beneath the water's surface, without setting off a ripple, and remains mostly submerged, with just its head and neck above water. Like a loon, it can also swim fair distances underwater. The young spend their first few weeks riding around on the backs of their parents, which even dive with chicks aboard. Grebes are designed for swimming; awkward on land, they rarely come ashore.

Identification 12–15″. Stocky, compact; brownish, with conspicuously short, chickenlike bill. Breeding birds have black band around throat and black ring around bill. Juveniles striped, with recognizable grebe bill.

Voice A low, hollow series of *cow-oo*, *cow-oo* notes.

Habitat Lakes, ponds, and marshes.

Range Breeds throughout most of North America south of central Canada. Winters from British Columbia and New England south.

34

Common Loon *Gavia immer*

Experts believe that the loons are representatives of a very primitive group of water birds; the earliest loon fossils date from 65 million years ago, and present-day loons have no close relative among modern birds. With their heavy, solid bones, loons dive easily and can stay underwater for two to three minutes, although most dives are much shorter. On still nights, lake shores of the North ring with this loon's hair-raising laughing call.

Identification 28–35". Large, ducklike. Breeding birds have black-and-white checkerboard pattern above, with velvety black head, white necklace, white underparts, and straight, black, daggerlike bill. Winter birds and immatures blackish above, whitish below, with paler bill.

Voice Loud, yodeling call given on breeding grounds; wild, laughing call given at night.

Habitat North country lakes and rivers; also in protected bays and estuaries, and out to sea during migration.

Range Breeds from Alaska to British Columbia and east to New England; winters south along both coasts.

Common Goldeneye *Bucephala clangula*

The Common Goldeneye can even be identified at night, for its wings produce a loud, distinctive whistling sound that has given rise to this duck's nickname, "the whistler." This species is famous for its spring courtship display: the drake stretches his head forward and then throws it back, uttering a loud, rasping noise, as the female obligingly looks on.

Identification 16–20″. Large and stocky. Breeding drake white with black back, dark, glossy, greenish-black head, and large round white spot in front of eye. Females, juveniles, and nonbreeding drakes grayish with brown head, and white underparts. Both sexes have black wings with white wing patch conspicuous in flight, and distinctive, puffy-looking head shape.

Voice Male gives nasal, rasping sound in courtship; otherwise silent.

Habitat Coastal bays and large lakes; also rivers and ponds.

Range Breeds from Alaska to Labrador, south to northern tier of states. Winters south to Florida and Mexico.

38

Greater Scaup *Aythya marila*

Also called the Bluebill, the Greater Scaup is often seen in winter on bays, estuaries, and large inland lakes, where it forms huge rafts, sometimes made up of thousands of birds. This species is closely related to the very similar Lesser Scaup *(A. affinis)*. But the Lesser Scaup breeds on the prairies of Canada, where it prefers small ponds and marshes, and it winters more widely across the southern half of the United States.

Identification 15–20″. Breeding drake has light gray mantle, dark breast, deep, glossy, blackish-green head and neck, and white belly; bill pale blue. Female dark brown above with gray-brown breast and white belly, and white patch around base of bill.

Voice Courting drake gives soft whistles and cooing notes; females give a harsh, low *arr* or *kerr*. Often silent.

Habitat Large lakes and ponds; saltwater areas in winter.

Range Breeds from Alaska and N. Canada to Maritime Provinces. Winters along both coasts and in Mississippi Valley.

40

Mallard *Anas platyrhynchos*

The Mallard is probably the best-known duck in the world. It was domesticated many centuries ago in Europe, and today wild Mallards still breed freely with domesticated white "puddle ducks." The Mallard has also been known to migrate great distances, and isolated populations have colonized remote places like the Hawaiian Islands.

Identification 18–27″. Breeding drake grayish, with green head, brown breast, and white neck-ring. Female and nonbreeding male mottled brown with white tail and orange-and-brown bill. Both sexes have glossy blue speculum, visible in flight.

Voice Female utters a loud, familiar quack; male gives a reedy *rah-rah-rah*.

Habitat Ponds, lakes, rivers, coastal marshes, bays, beaches, and city reservoirs.

Range Breeds from Alaska to Nova Scotia, south to N. Mexico, Texas, and Virginia. Winters from Washington to New York and south to Central America.

Green-winged Teal *Anas crecca*

These little dabbling ducks fly swiftly in compact flocks, wheeling and turning in unison. They often migrate in larger flocks than other marsh ducks; where food is abundant, it is not unusual to see several hundred birds traveling together. They feed in shallow marshes, mud flats, and flooded fields.

Identification 12–16″. Breeding drake has chestnut-colored head with bright green patch behind eye; vertical white stripe on side of breast; belly pale whitish; sides gray. Female, nonbreeding drake, and immature mottled gray and brown, with whitish belly. All have dark upper wings with green speculum.

Voice Drake has a clear, repeated whistle; female quacks.

Habitat Lake borders, ponds, marshes, mud flats, and wet fields.

Range Breeds from N. Alaska south to N. California, Colorado, Minnesota, and east across most of Canada to Newfoundland and New England. Winters from British Columbia and Utah south, east through Texas to Florida; also along Atlantic Coast from New York south.

44

American Black Duck *Anas rubripes*

The Black Duck is chiefly an eastern species and is rarely seen west of the Great Lakes. This duck appears black only at a distance; close at hand, it looks very much like a female Mallard; in fact, small groups of Black Ducks are often seen among large flocks of Mallards, and the two species regularly interbreed. Where they occur together, the Mallard can be distinguished by its orange-mottled bill and paler plumage.

Identification 19–22". Sooty brown with white wing linings, conspicuous in flight; head and neck slightly paler brown than rest of body; wings have purple speculum. Legs red, bill greenish. Sexes alike.

Voice Female gives a quack, almost indistinguishable from female Mallard's; male utters softer call notes.

Habitat Estuaries, marshes, mud flats, streams, and lakes.

Range Breeds from N. Alberta to Newfoundland, south to Minnesota and Virginia; winters from southern part of breeding range to northern Gulf Coast.

Canada Goose *Branta canadensis*

The Canada Goose is familiar and widespread throughout North America. These birds migrate in large, wavy V formations, piercing the air with their far-carrying, musical honk. Mated pairs stay together for many years, and their young remain with them until the family returns to the nesting grounds in spring. Birds always return to the same area to breed, and several distinct races have developed.

Identification	Large races 32–48″; small races 22–27″. Brownish-gray body and wings with long black neck and black head; distinctive white cheek patch usually extends under throat. Belly and flanks white. Bill and legs black.
Voice	Larger races give a resounding *onh-whonk;* small races cackle a high *ahnk.*
Habitat	Lakes, rivers, marshes, and bays; often in harvested cornfields and grasslands.
Range	Breeds from N. Alaska to Baffin Island, south to NE. California, Missouri, and Ohio. Winters from southern part of breeding range to Mexico and Gulf Coast.

Double-crested Cormorant *Phalacrocorax auritus*

This dark, long-necked water bird is often seen perched on a log or rock with its wings outspread; cormorants lack adequate waterproofing oils in their plumage, and they must periodically dry their wings in the sun. Cormorants are heavy-bodied diving birds that must paddle across the water's surface to gain momentum when taking flight.

Identification 30–36″. Adults black with bare patch of orange skin beneath bill; breeding adults have small tufts on each side of crown. Immatures brownish, paler on breast. Bill long, straight, with hooked tip in all plumages.

Voice Grunting calls near nesting colony; usually silent.

Habitat Lakes, rivers, and seashores.

Range In the East, breeds from Saskatchewan to Nebraska, eastward through Great Lakes region and Gulf of St. Lawrence; also along Atlantic, Pacific, and Gulf coasts. Winters from Long Island to Florida and Gulf Coast, and along Mississippi River and Rio Grande.

Brown Pelican *Pelecanus occidentalis*

This large bird is often seen in an attitude of repose, resting its head and huge bill on its breast. The prominent throat pouch allows the bird to scoop up a large mouthful of seawater and separate out the fish. The Brown Pelican performs spectacular dives, plunging from the air straight down into the water in pursuit of its prey. The American White Pelican, a larger relative, searches for fish as it swims through the water, often holding its wings open to block out the sun and reduce the glare from the water's surface.

Identification   45–54″. Adults stocky and dark brown with massive bill and throat pouch and whitish head. Breeding birds have cinnamon-brown on back of neck. Young birds have dull brown heads.

Voice   Adults silent; nestlings squeal and grunt.

Habitat   Bays, beaches, estuaries, and lagoons.

Range   In the East, breeds locally along coast from North Carolina to Florida; also on Pacific Coast. Occasionally wanders northward.

Great Egret *Casmerodius albus*

This large, magnificent heron is often seen in shallow water, where it stalks its prey slowly and deliberately. It feeds on a variety of aquatic and marsh-dwelling creatures, including fish, crayfish, frogs, and even snakes. At the turn of the century, fashion required that hats be decorated with plumes, and this species was nearly exterminated as a consequence of the unguarded enthusiasm of stylish ladies. It is now protected by law, but faces an equally serious threat to its existence as a result of the draining of its wetland habitat.

Identification 37–41″. Large, white; yellow bill, long black legs. Breeding adults sport long lacy white plumes on back.

Voice A low, guttural growl or croak.

Habitat Freshwater and saltwater marshes; marshy ponds or lake borders, lagoons, and tidal flats.

Range In the East, breeds from Minnesota and S. Ontario to Texas and Louisiana; along Atlantic from Long Island to Georgia and Florida; also in the Far West. Winters from North Carolina south, and in Texas and the Southwest.

Great Blue Heron *Ardea herodias*

The largest heron in North America, the Great Blue stands four feet tall and has a wingspan of seven feet. The "Great White Heron," for years believed to be a distinct species, is actually a form, or color phase, of this species. The Great Blue Heron is often seen standing perfectly still in shallow water, where it waits for fish or frogs to come within range of its long, sharp-pointed bill.

Identification 39–52″. Large, gray-blue, with long yellowish bill; head white, with black on crown and nape; legs dark. Immature similar to adult but paler, with all-black crown. "Great White Heron" all white with yellowish bill and legs.

Voice A harsh croak; usually silent.

Habitat Lakes, rivers, marshes, and ponds.

Range Breeds throughout United States and Canada, except for deserts and high mountains. Winters throughout most of southern half of United States, and along coasts north to New England and S. Alaska.

56

Turkey Vulture *Cathartes aura*

Also known as the "Turkey Buzzard," this large carrion-feeder is commonly seen sailing over open countryside or gliding along on shifting currents of air as it searches for food. In flight, the Turkey Vulture holds its wings in a slight V above the body, with its outer wing tips spread wide like the fingers of a hand. At dusk, Turkey Vultures often gather to roost in large numbers.

Identification 26–32″. Large and blackish, with small, unfeathered red head and stout bill with sharply hooked tip. Legs and feet orange. Silvery-gray wing linings conspicuous in flight, making wings appear two-toned. Immature has dark head and gray feet.

Voice Generally silent; utters hisses and groans at nest or when disturbed.

Habitat Dry, open country; often along roadsides; sometimes roosting in woods.

Range Breeds from S. British Columbia to S. New England, south to Mexico. Winters from New Jersey to Florida and E. Texas; also in parts of the Southwest.

58

Bald Eagle *Haliaeetus leucocephalus*

In 1782, when the Bald Eagle was chosen as the symbol of the United States, Benjamin Franklin protested. He considered this eagle "a bird of bad moral character," and recommended that the young nation adopt the Wild Turkey instead. The Bald Eagle once bred throughout North America, but its numbers have declined severely as a result of pesticide poisoning. Today some of these pesticides have been banned, and the species seems to be making a modest comeback.

Identification	30–43″. Very large, brown, hawklike bird with white head and tail and stout, hooked yellow bill. Immatures variable, but with dark head and tail and black bill.
Voice	A series of squeaky, thin cackling or chittering notes.
Habitat	Seacoasts, lakes, rivers, and marshes.
Range	Breeds in forested areas of Alaska and Canada south to Oregon, N. Idaho, Great Lakes area, and N. New England; also locally along Atlantic and Gulf coasts and in Florida. Winters from S. Canada south, especially along major river systems of the interior.

Osprey *Pandion haliaetus*

This large bird of prey is most often seen near water. It feeds chiefly on fish, which it captures with its large talons after plunging feetfirst into the water. Like the Bald Eagle, the Osprey has suffered a decline because it consumes fish contaminated with toxic chemicals. Ospreys build large, rough-looking stick nests in dead trees, and they are quick to take advantage of man-made platforms and telephone poles for their nesting needs.

Identification 22–25″. Large, hawklike; brown above, white below, with white head; dark brown line runs through eye and on side of face. Juveniles similar but more mottled. In flight, wings show distinctive bend at "wrist."

Voice Loud whistling and chirping given at nesting and during courtship; also a *kip kip ki-yeuk*, *ki-yeuk* when alarmed.

Habitat Coastal areas, lakes, and rivers.

Range Breeds from Alaska and north-central Canada to Newfoundland, south to California and Arizona, Great Lakes area, and Nova Scotia; south along Atlantic Coast to Florida and Gulf Coast. Winters on southern coasts.

Northern Harrier *Circus cyaneus*

Formerly known as the Marsh Hawk, this species is the only North American representative of the harriers, a large, diverse group of mainly Old World hawks. The Northern Harrier has very keen hearing, augmented by its disk-shaped facial feathers; it hunts on the wing, with its ears attuned to detect the slightest rustle in the grass or marshland below.

Identification 16–24″. Slim, with long wings and tail. Male light gray above, whitish with small reddish flecks below; tail obscurely barred. Female brown above with brownish streaks below. Both sexes have prominent white rump. Immature brown above, rusty below.

Voice Usually silent; a chattering *kee-kee-kee* near nest.

Habitat Grasslands, marshes, and open fields.

Range Breeds from Alaska to N. Alberta and east to Newfoundland; south to S. California, N. New Mexico, Ohio, and Virginia. Winters from Washington, N. Utah, Great Lakes region, and New England south to Mexico and Florida.

Red-tailed Hawk *Buteo jamaicensis*

The Red-tailed Hawk is the most abundant large hawk throughout much of North America; however, it is highly variable in color, and there are several distinct subspecies that may be difficult to identify. In general, the Red-tail appears stocky and broad-winged and usually has a dark head. Unlike the similar Broad-winged and Swainson's hawks *(B. platypterus* and *B. swainsoni)*, the Red-tail does not usually fly in large flocks.

Identification 19–25″. Large; typical bird dark brown above, usually light below with dark band on belly. Tail rufous with dark band and paler tip. Geographical variations include a dark brown color phase and a pale phase with a white tail. Immature has grayish tail with narrow bands.

Voice A loud, harsh, descending *tseer*, given when disturbed.

Habitat Grasslands, pastures, open woods, and farmlands; also in plains, tundra, and deserts with scattered trees.

Range Breeds throughout most of North America; winters from Canadian border south.

66

American Kestrel *Falco sparverius*

Formerly known as the Sparrow Hawk, the American Kestrel is the smallest North American falcon. Widely distributed, it is seen hunting from perches or on the wing; it sometimes hovers in the air before descending in a swift, stooping dive to take its prey. The American Kestrel has proved adaptable in the face of development, and it even nests in suburbs and big cities.

Identification 7½–8″. Small, with long, pointed wings and rusty tail and back. Adult male has blue-gray wings and rusty crown; female has black-barred rufous wings. Underparts white or buff, male's with black spots, female's with heavy streaks.

Voice A loud, shrill, *killy-killy-killy*.

Habitat Open countryside, grasslands, farms, suburbs, and city parks.

Range S. Alaska to Newfoundland, south through South America. Winters north as far as S. British Columbia, Illinois, and New England.

Common Nighthawk *Chordeiles minor*

In the East, the Common Nighthawk often nests in settled areas, and seems to prefer flat, gravelly rooftops. At the start of the mating season, male nighthawks flutter over the rooftops, giving their buzzy *bzeerp* or *brrrrrp* call; then they dive straight down, and as they plummet, the air whistling through their wings creates a sudden, explosive boom.

Identification   8½–10″. Mottled gray, white, black, and brown above; underparts buff with brown bars. Long, pointed wings marked with white patch near bend, visible in flight, as is white throat patch. Tail long. Female slightly duller than male.

Voice   A nasal, insectlike *beeerp* or *brrrrrrp*.

Habitat   Open woodlands, forests, meadows, sagebrush plains, and cities.

Range   Breeds from SE. Alaska east to Quebec, south to N. California, Nevada, SE. New Mexico, Texas, and Florida. Winters in the tropics.

Common Barn-Owl *Tyto alba*

Because of some unusual skeletal features, the Common Barn-Owl has been placed in a separate family from other North American owls. Like the typical owls, it has extremely sensitive hearing. The Common Barn-Owl, in fact, was used to prove that owls can locate prey in total darkness, solely by the use of their hearing.

Identification 14″. Golden brown above, with grayish mottling on wings and back; pale buff to white below, with sparse dark spots on breast and wing linings. Face heart-shaped, white, with long, narrow bill. Feet and legs covered with bristly white feathers.

Voice Song a long, rasping screech, increasing in volume, and a loud hiss.

Habitat Farm areas, marshes, prairies, and open woodlands; also suburbs and cities.

Range SW. British Columbia, South Dakota, N. Illinois, and S. New England south to Central and South America. Some northern populations move south in winter.

72

Great Horned Owl *Bubo virginianus*

This powerful nocturnal bird of prey is the largest North American "eared" owl. Its diet includes large animals such as skunks, ducks, opossums, and even hawks. Like other owls, the Great Horned has keen night vision and acute hearing that is enhanced by facial disks—feathers that encircle the eyes and direct sound to their ears, enabling the bird to hear the rustle of a mouse in a corn field. This adaptable owl is found in almost any habitat; it is often seen perched on dead trees or telephone poles in the early evening, searching for prey.

Identification   18–25″. Large with widely spaced ear tufts. Dark gray-brown with fine whitish mottling above; buff-white below, with dark brown barring and white throat. Eyes bright yellow.

Voice   A deep, sonorous, resonant series of hoots: *hoo, hoo-hoo-hoo, hoo, hoo;* or *hooo, hoo-hoo, hoooo, hoooo.*

Habitat   Forests, open country, swamps, deserts, and even large city parks.

Range   Throughout North America; usually does not migrate.

74

Eastern Screech-Owl *Otus asio*

This little owl of the East can be recognized by its ear tufts; other small owls in its range are "earless." Despite its size, the Eastern Screech-Owl can be aggressive, and it has been known to strike people on the head when they unwittingly approach its nest. By day, Eastern Screech-Owls often assume a frozen, upright posture, relying on their mottled plumage for camouflage. This bird and the nearly identical Western Screech-Owl were formerly believed to be a single species. Their ranges overlap only in a small area of Texas, where they must be distinguished by voice.

Identification 7–10″. Small with mottled upperparts and prominent ear tufts. Whitish below, with streaks and bars. Two color phases: grayish and rufous, with brownish intermediates.

Voice A long, tremulous, descending whinny; also short barks and soft purrs.

Habitat Deciduous forests, woodlands, suburbs, and farms.

Range N. Minnesota and adjacent Canada east to N. New England and south to Texas and Florida.

Ring-necked Pheasant *Phasianus colchicus*

Introduced from Asia, the Ring-necked Pheasant has taken to its adopted home so well that it has become our most adundant upland gamebird. With its colorful plumage and overall chickenlike appearance, this bird is easy to recognize. Very tolerant of man, it is often found near towns and cities. This pheasant is a fast runner; when it flies, it takes off with a startling flurry of sound.

Identification
Male 30–36"; female 20–26". Chickenlike, with long, pointed tail. Male has bright green or blue-green head, red face, white neck-ring; body and wings iridescent bronze, gold, and red, with bold, dark spotting. Female soft brown, spotted and barred with black.

Voice
A loud, crowing *cuck-cuck* or *caw-caw*, accompanied by loud wingbeats. Males cackle when taking off.

Habitat
Grassy and brushy areas near woodlands; farms, pastures; also in cattail marshes in winter.

Range
Throughout much of United States and S. Canada in suitable agricultural regions; absent from high mountains, deserts, and most of the Southeast.

78

Northern Bobwhite *Colinus virginianus*

The cheerful, whistled "bob-white!" call of this quail is a familiar sound of summertime. It is the male that gives this call; once he has successfully attracted a female, he stops calling and devotes his energy to family life, sometimes even helping to incubate the eggs. After the breeding season, families form coveys of 8 to 25 birds, which remain together in the cover of low bushes and often huddle together for warmth on cool evenings.

Identification 8–11". Small, plump; chestnut or reddish brown above, white below with black mottling; sides streaked with chestnut. Throat and eye stripe white in male, buffy in female; both with dark eyeline and crown.

Voice Male gives a clear, rising *bob-white* or *poor-bob-white;* both sexes give a repeated, whistled *ka-whoi-lee.*

Habitat Open areas with brush; farmlands, pastures, and forest edges.

Range Nebraska and Iowa south to Texas and east to New England and Florida; introduced in Washington, Oregon, and Idaho.

Mourning Dove *Zenaida macroura*

The soft, melancholy call of this bird, which gave rise to its common name, is most often heard at first light in spring and summer. The Mourning Dove is one of our most common and widespread birds, and it is often seen in large flocks. It is hunted extensively in some areas but protected in others.

Identification 11–13″. Slim, with small head and long, tapered tail. Soft, sandy brown or brownish gray above, with a few black spots; paler below, sometimes washed with pale cinnamon; tail feathers tipped with white.

Voice A low, sad *whoo-oo, hoo, hoo, hoo;* second note rises sharply.

Habitat Almost anywhere except dense forests: woodlands, streamsides, desert washes, gardens, city parks, and suburban backyards.

Range Breeds from S. Alaska and W. British Columbia through S. Alberta to Great Lakes region to New England; south to Mexico and Florida. Northern populations migratory.

82

Rock Dove *Columba livia*

Everyone recognizes the Rock Dove, commonly called the Pigeon. Introduced from Europe, this ubiquitous bird was originally a denizen of the rocky sea cliffs high above the Atlantic and Mediterranean. In the wild, these birds eat grass seeds, which they obtain by taking a head of grass in the bill and shaking it sharply to break some seeds loose. In cities, where they subsist almost entirely on bread crusts, bits of discarded pizza, and other baked goods, this instinctive feeding technique persists.

Identification 12–13″. Stocky, with short fan-shaped tail. Birds in the wild (and many in cities) bluish-gray with 2 narrow black wing bars, white rump, and some iridescence on sides of neck. Color variations include all-black, all-white, piebald, and reddish-brown birds.

Voice A soft, rolling *coo-croo* or *coo-took-crooo*.

Habitat Cities, parks, gardens, suburbs, and farm areas; rocky canyons or sea cliffs.

Range Throughout S. Canada and United States; does not migrate.

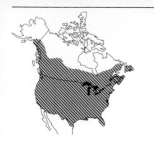

European Starling *Sturnus vulgaris*

Like the Rock Dove, the Starling is native to Europe. In 1890, a flock of 100 of these birds was released in New York City's Central Park; since that time, this aggressive species has expanded its range to include most of the continent. Starlings are unpopular in many areas because they damage fruit crops and compete with native species for nesting cavities, but they have proven efficient in helping to reduce the population of insect pests, such as locusts.

Identification	7½–8½″. Short-tailed, chunky. In spring, black with iridescent greenish gloss; bill yellow. Winter plumage heavily flecked with white; bill dark. Immatures dusky gray-brown above, paler below.
Voice	A wide variety of squeaks, chattering notes, whistles, and clicks; often gives a "wolf-whistle" and mimics other birds.
Habitat	Cities, parks, orchards, woodlands, and farmland.
Range	Throughout the United States and S. Canada

Common Grackle *Quiscalus quiscula*

Abundant in residential areas, the Common Grackle is handsome, but it has earned a reputation as a disturber of the peace. Grackles are especially noisy in winter, when they gather in huge roosts with equally vociferous blackbirds and European Starlings. There are two color forms of the Common Grackle, bronze and purple; the latter occurs only in the Southeast and along the Atlantic Coast.

Identification | 11–13½". Adult appears all black from a distance. Glossy purple or bronze tones visible at closer range; head has bluish sheen. Tail long, wedge-shaped, appearing keeled in flight. Eyes yellow. Juvenile bird dark brownish, with dark eyes.

Voice | A loud, clucking *tschaak;* also a high, ascending screech.

Habitat | Open woodlands, parks, gardens, and lawns.

Range | Breeds east of Rockies, from NE. Alberta to Nova Scotia and south to Texas and Florida. Winters in southern two-thirds of range.

Brown-headed Cowbird *Molothrus ater*

Cowbirds are brood parasites: They do not raise their own young, but lay their eggs in the nest of another species, after having removed one egg from the host's clutch. When it hatches and begins to grow, the young cowbird is usually so much larger than the young of its foster parents that the other nestlings either starve or are crowded out. Despite their limited family life (for these birds form only brief pair bonds), the male performs an elaborate courtship display, embellishing the show with loud whistles and bubbling notes.

Identification   6–8″. Male iridescent greenish black with deep brown head; female gray-brown; juvenile gray-brown with faint streaks on breast and scaly-looking upperparts. Bill black and finchlike in all plumages.

Voice   Voice squeaky, bubbly, high-pitched; female chatters.

Habitat   Woodlands, farmlands, fields, and suburbs.

Range   Breeds throughout most of the United States and S. Canada, extending into N. Alberta. Winters in S. United States.

90

American Crow *Corvus brachyrhynchos*

Crows are wary birds, and opportunistic in the face of changes—with the result that the species is far more abundant today than it was before the arrival of European settlers to this country. For all its intelligence, the American Crow is widely regarded as a nuisance; this judgment is perhaps unfair, because crows are useful in keeping down the populations of insects that damage crops. Crows sometimes gather in roosts of several hundred thousand birds.

Identification	17–21″. Large and stocky. Black all over, with slight purplish sheen. Bill stout; tail squared or fan-shaped.
Voice	A raucous, familiar *caw, caw, caw*.
Habitat	Open areas, woodlands, fields, suburbs, orchards, gardens, and city parks; tends to avoid deserts and dense forests.
Range	Breeds throughout southern two-thirds of Canada and most of United States; winters south of Canadian border. Absent from much of interior Southwest.

Red-winged Blackbird *Agelaius phoeniceus*

This pretty black bird with red epaulets is common and well known. It usually nests in wetlands, and along the East Coast it can be easily seen through the window of any train traveling between northern Connecticut and Philadelphia, in the marshy areas that border much of the railbed. Redwings congregate with other blackbirds after the breeding season, often forming flocks of many thousand birds and roaming over open country.

Identification 7½–9½". Male black with bright red shoulder patches. Female and juveniles have heavy, dusky brown streaks.

Voice Song a liquid, musical *oh-ka-lee!* Also various *chuck* and *kink* notes.

Habitat Usually nests in marshes and other wetlands, especially areas with cattails; also in moist thickets, pastures, and meadows.

Range Breeds from S. Alaska, N. Alberta, and Ontario to Maritime Provinces, south through entire United States. Winters in most of southern two-thirds of United States and the temperate Northwest.

94

Northern Cardinal *Cardinalis cardinalis*

The Cardinal is a favorite visitor to backyards and city parks, beloved for both its cheerful song and its brilliant red plumage. This species is a seed-eater; its sharp-edged, conical bill is perfectly suited to cracking open nuts to get at the kernel within. Cardinals may raise as many as four broods of young each year; they often nest in suburban areas, where their song is heard all year.

Identification 7½–9″. Male brilliant red with backswept crest and black face; bill red. Female grayish-olive above, paler below, with varying amounts of red on crest, wings, and tail; bill pink; crest and black area around bill smaller than in male. Immature gray, washed with buff below, with dark bill.

Voice Song a whistled *what-chew, what-chew, wit wit wit;* also *birdy-birdy-birdy;* many variations.

Habitat Woodlands, gardens, thickets, and areas with brushy undergrowth.

Range The Dakotas to Nova Scotia, south to Texas and Gulf Coast; also in Arizona and S. California; nonmigratory.

Rose-breasted Grosbeak *Pheucticus ludovicianus*

This pretty black-and-white bird with its bright blaze of red is often conspicuous in early spring before the trees leaf out. The males deliver a musical, warbling song; surprisingly, they often sing from the nest while tending the eggs. These birds sometimes produce two broods a year; the male feeds the first brood of nestlings while the female prepares the second nest.

Identification 7–8½″. Male black above, white below, with white wing patches and bright splash of rose-red on breast; underwings also red. Female streaky dark brown above, grayish below with heavy brown streaks, prominent white eyebrow; wing linings yellow. Bill stout and pale.

Voice Song similar to American Robin's but softer: *cheerily, cheer-up, cheerily.* Also a distinctive *chink* or *clink* note.

Habitat Open areas near moist woodlands or thick shrubs; also old orchards.

Range Breeds from Alberta, Ontario, and Great Lakes east to Nova Scotia, south to Kansas and New Jersey; also south to N. Georgia in Appalachians. Winters in tropics.

Cedar Waxwing *Bombycilla cedrorum*

The Cedar Waxwing is an elegant bird, with markings so neat they seem almost to be painted on. These birds travel in noisy flocks; they visit orchards and garden trees to feed on berries or flower petals, often departing with the suddenness of decamping gypsies. Cedar Waxwings can be seen perched in a row along the branch of a fruit tree, passing berries down the line from one bird to the next.

Identification 6½–8″. Sleek bird with soft brown upperparts and breast, and bold, black mask through eyes; wings gray with hard, red, waxy tips on inner feathers; belly yellowish; tail dove-gray with yellow tip.

Voice A thin, high, lisped *ssseee* or *tseee tseee tseeee*.

Habitat Orchards, residential areas, and open woodlands; especially with fruit trees.

Range Breeds from SE. Alaska and British Columbia to Newfoundland, south to N. California, Virginia, and in mountains to Georgia. Winters through most of United States, but absent from high mountains.

100

Eastern Meadowlark *Sturnella magna*

Although it superficially resembles the true larks, a chiefly Old World group, this bird is not a lark at all, but a member of the blackbird family. The Eastern Meadowlark is an enthusiastic vocalist, delivering its cheery song from a prominent perch. These birds are most often seen in open, grassy areas and in farm country. When alarmed, they flash their tails, revealing conspicuous white outer tail feathers.

Identification 7–10″. Stocky; upperparts patterned in streaky buff, black and brown; throat and breast bright yellow, with bold black V on breast; tail has white outer feathers. Bill straight and pointed.

Voice Clear, loud whistle, *see-you, tee-year;* often in series. Also gives a harsh alarm note.

Habitat Grasslands, farmlands, meadows; also seen along roads.

Range Breeds from S. Manitoba to Nova Scotia, south to Texas and Florida; also in extreme W. Texas, New Mexico, and S. Arizona. Winters in all but northernmost part of breeding range.

102

Yellow Warbler *Dendroica petechia*

Common and widely distributed, the Yellow Warbler has been described as "a bit of feathered sunshine." This bird is often the unwilling host of cowbirds, which lay their eggs in the nests of other birds. The Yellow Warbler fights back against the cowbird by covering the interloper's egg with another layer of nest material to prevent its hatching; a few Yellow Warbler nests have been discovered with as many as six layers.

Identification 4½–5″. Bright yellow below, yellow-green above, with 2 bold yellow patches in tail; male has thin rusty streaks on breast, and sometimes a rusty crown (in Florida Keys); usually somewhat brighter than female. Immature female olive-green; immature male resembles female.

Voice A cheerful, musical *sweet-sweet-sweet, sitta, sitta, see;* also a soft, distinctive *chip* call.

Habitat Woodlands and thickets, especially near streams; often in alder and willow thickets, gardens, and swampy areas.

Range Breeds from Alaska through most of Canada and United States. Winters in the tropics.

American Goldfinch *Carduelis tristis*

Goldfinches feed mainly on seeds, especially those of thistles, goldenrods, and other weedy plants. This food is not abundant until the summer is fairly well advanced, so the American Goldfinch must put off its nesting season until then. Using plant fibers, this bird weaves nests that are almost watertight; in heavy rain, nestlings not protected by their parents have been known to drown.

Identification 4½–5½″. Breeding male bright yellow with black forehead, black tail, and black on edge of wings; rump and wing bars white. Female and nonbreeding male duller, with some gray or greenish; tail and wings black; wings have white bars; lack black on forehead.

Voice Song a bright, sweet *per-chick-o-ree*.

Habitat Thickets, streamside willows, weedy grasslands, and deciduous trees; avoids treeless areas and dense forests.

Range Breeds from SE. British Columbia and N. Alberta to Newfoundland, south to S. California, Utah, Nebraska, Oklahoma, and S. Carolina. Winters from most of breeding range south to Mexico.

106

Yellow-rumped Warbler *Dendroica coronata*

Known formerly as the Myrtle Warbler, this bird has been shown to be the same species as a western bird, Audubon's Warbler. Both have been officially combined under the new name, Yellow-rumped Warbler, but many people prefer to use the old names for clarity. The two forms intergrade only in a very narrow area in British Columbia.

Identification 5–6″. Breeding male slate-blue above with darker streaks; rump bright yellow; charcoal-gray patch around eye; throat white; crown and sides have small, bright yellow patch. Female, immature, and nonbreeding male mainly gray-brown, echoing pattern of breeding male, but duller; with yellow rump and white spots in tail.

Voice A thin, musical trill, *twee-twee-twee;* also a sharp *chek!*

Habitat Mixed and coniferous forests.

Range The Myrtle breeds from Alaska to Newfoundland, south through most of Canada and New England; winters chiefly in the Southeast. Audubon's breeds and winters in the West; ranges overlap only in British Columbia.

108

Common Yellowthroat *Geothlypis trichas*

These abundant birds carefully conceal their nests in dense weeds on or near the ground. To thwart any potential predators, the Common Yellowthroat approaches the nest with great stealth, skulking through the weeds; after delivering the food, it leaves by another route. Yellowthroats are not at all secretive in courtship, however; the male gives a pretty show, flying high in the air and producing a jumble of notes.

Identification 4½–6". Greenish-brown or olive above with bright yellow breast and throat. Male has broad black mask across eyes; female and immature lack mask.

Voice A fast, repeated *witchity-witchity-witchity-witchity-wit*; call note a *tchip* or *chik*.

Habitat Brushy swamps, moist brambles and thickets, streamside growth, grassy marshes, and forest edges near water.

Range Breeds throughout most of North America; winters north to S. Carolina and central California.

110

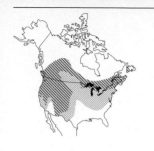

Evening Grosbeak *Coccothraustes vespertinus*

Despite their names—inspired by their similar large, conical bills—the Evening Grosbeak and the Rose-breasted Grosbeak are not closely related. The Evening Grosbeak is highly gregarious, occurring in large flocks; smaller flocks are seen during the breeding season. In flight, this bird can be recognized by its white wing patches and loud call notes.

Identification 7–8½". Stocky, with large, greenish-yellow, conical bill. Male has brown head with bold yellow forehead; back and belly bright yellow; wings black with large white patches. Female much grayer, with little yellow.

Voice Song a wandering series of musical whistles; call note a loud, ringing *cleep*.

Habitat Coniferous and mixed forests; a variety of habitats after breeding season.

Range Breeds from British Columbia to Nova Scotia, south to N. New England, Minnesota, and throughout western mountains. Winters south of breeding range through much of the Midwest and Middle Atlantic States.

112

Northern Oriole *Icterus galbula*

The eastern form of this species is widely known as the Baltimore Oriole. This bird weaves a hanging, pouchlike nest of plant fibers and bark, and readily accepts string or yarn. The Baltimore Oriole's bright orange-and-black plumage echoes the colors on the coat of arms of George Calvert, the first Lord Baltimore, who was granted the province of Maryland by James I in 1632.

Identification 7–8″. Male has black hood, upper back, and wings, and bright orange breast, belly, and lower back; wings have white patches. Female olive to greenish-gray, vaguely streaked and washed with light yellow. Immature male looks like female but more orange, with dark hood and back gradually filling in.

Voice Song a flutelike, clear, varied whistle, given in short phrases; also chatters when alarmed.

Habitat Deciduous woodlands, residential areas, and shade trees.

Range Breeds from central Alberta to Nova Scotia, south to Oklahoma and Georgia. Winters mainly in tropics.

114

American Redstart *Setophaga ruticilla*

The name "redstart" comes from an old German word derived from the Anglo-Saxon word, *steort*, for "tail"; it was originally given to a rusty-tailed European thrush. The American Redstart is a lively, energetic little warbler that can be seen drooping its wings, fanning its bright orange-and-black tail from side to side, and darting out from a branch to catch flying insects. It is abundant through much of the East, and has extended its range with the expansion of second-growth forests.

Identification 4½–5½". Male glossy black with bright orange patches on wings and tail and white belly. Female and immatures olive-brown above, whitish below, with yellow patches on wing and tail.

Voice Song usually has 5 or 6 distinct, high notes, rising or falling at end: *chewy-chewy-chewy, chew, chew, chew.*

Habitat Open, second-growth woodlands and deciduous forests.

Range Breeds from SE. Alaska to Newfoundland, south to Utah, central Alberta, Oklahoma, and North Carolina. Winters from S. Texas and S. Florida south.

American Robin *Turdus migratorius*

Popularly regarded as the best sign of spring's arrival, the Robin is perhaps the most familiar songbird in all of North America. It is often seen running about on the ground, stopping every few feet to cock its head and watch carefully for signs of earthworms, its favorite food. In winter, these birds feed on berries and other fruits.

Identification 9–11". Dark gray-brown above, with brick-red or orange breast and belly; head and tail blackish, throat white. Juvenile gray-brown above, pale orange with black spots below.

Voice A rich, loud song of rising and falling phrases: *cheerily, cheerily, cheer-up, cheer-up*. Also a loud *weep* note and a lisped *see-lip* in flight.

Habitat Woodlands, forests, gardens, and suburban backyards.

Range Breeds throughout most of North America; winters mainly in southern two-thirds of United States; southernmost populations may be nonmigratory.

Rufous-sided Towhee *Pipilo erythrophthalmus*

Towhees often reveal their presence by noisily scratching in the underbrush; these birds kick both feet backward at once to uncover insects and seeds on the ground. The Rufous-sided Towhee has several races, or subspecies, in North America; in the East the sexes are markedly different, but in the West they are more nearly alike.

Identification 7–8½". Eastern male has black head, throat and upperparts; breast and belly white, bounded by rufous sides and flanks; white wing bars and tail spots visible in flight. Female similar, but black areas replaced by warm brown.

Voice Song variable, usually a trilled *drink-your-tea;* call note a slurred *to-whee?*

Habitat Woodlands, forest edges, gardens, and parks with low shrubby growth; avoids dense forests and treeless plains.

Range Breeds from S. British Columbia to Maine, and south to Mexico and Florida; absent from much of Great Plains. Winters throughout southern two-thirds of United States.

Gray Catbird *Dumetella carolinensis*

Although somewhat less accomplished than the related Mockingbird, the Gray Catbird is a skillful mimic; it imitates not only the songs of other birds, but also the utterances of barnyard hens and the croaks of tree frogs. The common name is a reference to the catlike squeal or whine that also forms part of this species' varied repertoire. Catbirds can be lured from shady cover by an imitation of this call.

Identification 9″. Slim, long-tailed. Dark gray above and below, with a black cap and rusty undertail coverts.

Voice Song a long, irregular jumble of notes on varying pitches; also a catlike *mew* or *meow*, and a variety of other notes.

Habitat Undergrowth, gardens, thickets, and residential areas; usually in dense cover.

Range British Columbia to Nova Scotia, south to Washington, N. Texas, and N. Florida. Winters along mid-Atlantic and Gulf coasts, south to Central America.

Purple Martin *Progne subis*

In the East, the Purple Martin has become so accustomed to the convenience of living in man-made martin houses that it rarely occupies any other abode. A cavity-nester, it originally dwelt in holes in tall dead trees and formed large colonies. Today, a large martin house may contain as many as 200 mated pairs of these birds. In late summer, Purple Martins congregate in huge roosts of up to several thousand individuals.

Identification   7–8¼". Adult male glossy blue-black overall. Female and immature male slightly duller above and pale gray below. Tail in all plumages long and shallowly forked.

Voice   Song a series of rich, gurgled notes; also a clear *tee-tee-tee*.

Habitat   Open woodland, farm country, and residential areas.

Range   Breeds from E. British Columbia and Manitoba east through Great Lakes region to Nova Scotia; south to New Mexico, Texas, and Florida; also along Pacific Coast. Winters in South America.

124

Barn Swallow *Hirundo rustica*

These graceful, abundant birds are strong fliers; they migrate great distances every spring and fall, and may cover as much as 600 miles in a single day. Skilled aerialists, swallows execute complicated dives and swoops, holding their large mouths wide open to catch insects on the wing. Barn Swallows nest chiefly in or on buildings, attaching their nests to walls and eaves.

Identification 5¾–7¾". Blue-black above with rusty or buff underparts (paler in female); tail very long and deeply forked, with white spots near base below. Forehead and throat usually a deeper rust than breast.

Voice A continuous series of twittering and chattering notes; also a liquid *slip-lip tsit-tsit*.

Habitat Open areas, farmland, marshes, lakeshores, and suburban areas; usually near water.

Range Breeds from Alaska, N. British Columbia, and N. Alberta through S. Canada to Nova Scotia, and south in most of United States; absent from Gulf Coast and parts of the Southeast. Winters in South America.

126

Eastern Bluebird *Sialia sialis*

The Eastern Bluebird is a colorful, cheerful herald of spring, and a favorite among songbirds. Bluebirds are the only North American members of the thrush family that nest in cavities; the Eastern Bluebird makes its home in old woodpecker holes, fence posts, and bird boxes. This species has experienced a decline in recent decades, possibly as a result of competition with starlings and House Sparrows for nesting sites.

Identification
6–6½". Male bright blue with chestnut throat and breast and white belly; female similar but with less brilliant color overall. Young bird has spotted breast.

Voice
Song a melodious warble. Call note a rich, liquid *queedle* or *chur-lee*.

Habitat
Open woodlands, farmland with scattered trees, and orchards.

Range
Breeds in S. Canada and United States from Great Plains to Atlantic Coast. Winters in southern half of breeding range.

Blue Jay *Cyanocitta cristata*

Handsome and aggressive, the unmistakable Blue Jay is common throughout the East. These birds are usually noisy, and often gather in flocks to scold owls, hawks, foxes, and other predators. But during the nesting season Blue Jays suddenly fall silent, as they stealthily visit their nests. Once their single brood has been raised the uproar starts again, and lasts through the fall and winter and into the following spring.

Identification 11–12½". Large; bright blue above with a prominent blue crest, black facial markings and bold black-and-white pattern in wings; tail long, with horizontal black bands and white tips. Underparts grayish white.

Voice A harsh, noisy *jay-jay;* also a loud *thieef-thieef,* a musical *queedle,* and a variety of other calls.

Habitat Widespread; particularly common near oak or pine forests.

Range East of Rockies from S. Canada to Oklahoma, E. Texas and Florida. Northern birds usually move south in winter.

Belted Kingfisher *Ceryle alcyon*

The long, daggerlike bill and backswept crest of this bird give it a large-headed appearance. The Belted Kingfisher is usually solitary; it stakes out a favorite section of a stream or lakeshore and sits conspicuously on an overhanging limb, waiting to spot a fish. Then, with a sharp rattling noise, it plunges into the water, coming up with its catch in the bill. These birds make their home in a cavelike hollow, which they excavate in a steep riverbank or seaside cliff.

Identification 12–14″. Blue-gray above, white below, with long, daggerlike bill and bushy crest; white collar about neck and broad blue-gray band across breast; female also has rufous band crossing upper abdomen.

Voice A loud, long, dry rattle.

Habitat Rivers, lakes, and seashores.

Range Breeds in much of Alaska and S. Canada, south through United States; absent from deserts of the Southwest. Winters along coasts and from Great Lakes through Mississippi Valley; also farther south.

Eastern Kingbird *Tyrannus tyrannus*

Like other tyrant-flycatchers, the Eastern Kingbird feeds principally on insects, which it captures on the wing; at some seasons of the year it also takes berries and small fruits, plucking them as it flies with shallow wingbeats. The Eastern Kingbird is conspicuous and noisy, and calls frequently in flight.

Identification	8½–9½″. Dark bluish-black above, especially on crown and sides of face; white below. Tail long and black, with white tip.
Voice	A harsh *kip-kip-kipper-kipper* or *killy killy killy;* also chatters and screams.
Habitat	Woodland edges, country roadsides, farms, streamsides, and orchards.
Range	Breeds from N. British Columbia east to Maritime Provinces, south to NE. California, central Texas, Gulf Coast, and Florida. Winters in South America.

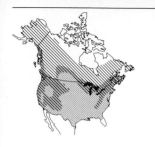

Dark-eyed Junco *Junco hyemalis*

This species includes five forms that were until recently believed to be separate species. The only one that occurs regularly in the East is commonly called the Slate-colored Junco. It is a lively bird, often seen hopping about on the ground near shrubby areas, searching for insects and seeds. When disturbed, it flies up into the branches, showing its white outer tail feathers. Juncos are common visitors to feeders in winter throughout much of the East.

Identification 5–6½″. Slate-gray above and on breast, with white belly, white outer tail feathers, dark eyes, and pink bill; female and some fall males may have back washed with buff.

Voice A tinkling trill, almost always on one pitch. Also some twittering and a *tick* alarm note.

Habitat Woodlands, fields, brushy areas, and lawns.

Range Breeds from Alaska to Newfoundland, south to Manitoba and New England, and south in mountains to Georgia. Winters from S. Canada to Florida and Texas, occasionally in S. California. Other forms in West.

Eastern Phoebe *Sayornis phoebe*

The Eastern Phoebe sings out its name, *fee-bee*, *fee-bee*, as it pumps its tail up and down. This bird belongs to a large family called the tyrant-flycatchers (because many family members attack hawks and other large birds that approach their nests). Flycatchers are mainly insect-eaters: From a suitable station on a branch or ledge, they dart swiftly out to snatch a flying insect, then make a quick reverse and fly back to their perch.

Identification 6½–7". Blackish or olive above, darkest on head; paler below, with long tail. Lacks wing bars and eye-ring (present in similar species).

Voice Song a repeated *fee-bee*, with second note higher or lower than first. Call note a sweet, short *chip*.

Habitat Open woodlands, cliffs, and building ledges; almost always near streams, brooks, or rivers.

Range Breeds east of Rockies from S. Mackenzie and N. Saskatchewan to Texas, east to north-central Quebec, south to Texas and South Carolina. Winters from S. Oklahoma, Gulf Coast, and Virginia south to Mexico.

Tufted Titmouse *Parus bicolor*

Vocal and common, the Tufted Titmouse is often seen scrambling up and down the bark of trees—and sometimes hanging upside down—in its energetic pursuit of insects. After the breeding season, Tufted Titmice join groups of other little birds, chiefly chickadees, nuthatches, and Downy Woodpeckers; these small bands of birds rove together, often visiting feeders en masse.

Identification	4½–5½″. Gray above, with white or pale gray underparts and reddish sides; conspicuous dark gray crest (black in some Texas birds).
Voice	Song a repeated, whistled series of *peter-peter* notes. Also a lisping call.
Habitat	Moist woodlands and brushy swamps; also residential areas, towns, and city parks in winter, often at feeders.
Range	S. Minnesota and Wisconsin to Maine, south to Florida and Gulf Coast; does not migrate.

140

Red-breasted Nuthatch *Sitta canadensis*

Nuthatches are plump-looking, energetic little birds that often forage on the smaller twigs and branches of trees. Their quest takes them to tree trunks as well, where they can often be seen making their way headfirst in a spiral path from the upper part of the trunk to the base. The Red-breasted Nuthatch inhabits coniferous forests, moving in a desultory fashion from one tree to the next.

Identification 4½–4¾″. Blue-gray above, rusty below, with paler throat; face and eyebrow white with bold black eye stripe; crown and nape black in male, grayish in female.

Voice A nasal, tinny *yank-yank* or *nyak-nyak*.

Habitat Coniferous forests; often ranging widely in winter to other habitats if conifer seeds are scarce.

Range Breeds from SE. Alaska to Newfoundland, south to S. California, Colorado, Michigan, and New Jersey; south in mountains to North Carolina. Winters in most of breeding range and south to Mexico and Gulf Coast.

Black-capped Chickadee *Parus atricapillus*

Flocks of little chickadees are a frequent sight at feeders in winter. These birds are brave and curious, and can easily be persuaded to take food from a person's hand. The Black-capped Chickadee is a close relative of the Tufted Titmouse, and the two species are often found together in small groups after the breeding season. The Black-capped Chickadee is the state bird of Massachusetts and Maine.

Identification 5¾". Gray above, with black forehead and cap, and black throat; paler gray below and over ears, with white cheeks. Wings have narrow white edges.

Voice Usual call a buzzy *chick-a-dee-dee-dee;* song a mellow, whistled *fee-bee,* with second note lower.

Habitat Mixed and deciduous forests, open woodlands; often in residential areas, especially in winter.

Range Breeds from Alaska to Newfoundland, south to N. California, Missouri, and N. New Jersey. Winters in much of breeding range, also south to Oklahoma and Maryland.

144

Golden-crowned Kinglet *Regulus satrapa*

The Golden-crowned Kinglet is a tiny, very round, and very energetic bird. It often travels in the company of its relative, the Ruby-crowned Kinglet (*R. calendula*), which is similar but has a red tuft on the top of its head and a white eye-ring. Both kinglets nervously flick their wings as they sit on a twig—a help in identification.

Identification 3½–4″. Olive above with 2 white wing bars; paler grayish white below; white eyebrow; bill slender, tail has small notch. Crown deep gold or orange-yellow in male, yellow in female, and bordered with black in both sexes.

Voice A thin, rising *see-see-see;* also a *tseeep.*

Habitat Conifer forests in nesting season; other woodland habitats as well in winter.

Range Breeds from Alaska to Newfoundland, south in suitable conifer-forest zones to California, Colorado, Minnesota, and southern Appalachians; absent from Great Plains and most of the Southeast. Winters from southernmost Canada throughout most of United States.

146

White-throated Sparrow *Zonotrichia albicollis*

This sparrow is one of the commonest migrants in much of the eastern United States; from deep within a thicket, members of a flock announce their presence with a thin, sweet call note. On warmer days, the Whitethroat sings a sweet, mournful song that is easy to recognize; a whistled imitation often lures these birds from cover.

Identification 6½–7¼". Brown with black streaks above; clear gray below. Head striped in bold black-and-white, with white patch on throat and yellow splotches between bill and eye. Sexes similar, but female and some males usually duller.

Voice Song a whistled, sad, *poor Sam Peabody, Peabody, Peabody, Peabody*. Also gives a *tseet* flocking call.

Habitat Conifer forests with brushy undergrowth; woodlands, brushy pastures, and residential areas in winter.

Range Breeds from Mackenzie to north-central Quebec and Newfoundland, south to Minnesota, Pennsylvania, and Massachusetts. Winters from southern part of breeding range to Gulf Coast and Mexico.

148

Red-eyed Vireo *Vireo olivaceus*

Throughout its breeding season, this species fills the forests with lively music, singing from sunrise to twilight, and even sometimes continuing to sing as it captures large insects. On very hot days, it may be the only bird singing. Despite its continuous vocalizations, however, the Red-eyed Vireo can be difficult to observe; less energetic than many other birds, it keeps a low profile amid the trees.

Identification   5½–6½". Olive above, whitish below, with red eyes visible from a short distance; white eyebrow bordered with black stripe above; eye also has black stripe. Crown gray. Lacks wing bars.

Voice   Song a continuous series of short, musical phrases and brief pauses, repeated in sequence: *tee-yew, chew-wee; cheerio, ter-wip, tee-yew.*

Habitat   Deciduous forests and residential areas with shade trees.

Range   Breeds from N. British Columbia through most of S. Canada, south to Oregon, Colorado, Texas, and Florida. Winters in South America.

150

Chipping Sparrow *Spizella passerina*

Abundant and familiar, the Chipping Sparrow has become a common bird in suburban gardens and residential areas. It sings frequently, but often remains inconspicuous. Chipping Sparrows are opportunistic in their approach to nest-building; they seem to prefer horsehair as a lining material, but will take hair from cattle, bison, and even dogs.

Identification 5–5¾". Adult has black forehead, rusty crown, white eyebrow, and black eye-stripe. Upperparts streaked in brown and black, with 2 white wing bars; underparts, cheek, and back of neck clear gray. Immature buffier, more streaked, without bold black-and-white markings or rusty crown.

Voice A thin, insectlike trill on 1 note. Call a sweet, high *tseep*.

Habitat Forest edges, orchards, city parks, and gardens.

Range Breeds throughout most of Canada and south throughout United States to S. Arizona, New Mexico, and S. Texas; absent from most of Florida. Winters from S. California, S. Texas, and Maryland southward.

Song Sparrow *Melospiza melodia*

The Song Sparrow is one of the best known and most widespread songbirds in North America. It displays much geographical variation, and there are 34 recognized subspecies, varying from very pale to very dark; some isolated races are also much larger than their more familiar counterparts in the rest of the continent. The Song Sparrow pumps its tail in flight.

Identification
5¾–7″. Brown above with grayish streaks; white below with heavy brown streaking and large spot at center of breast. Tail usually has more reddish brown than back.

Voice
Song has 3 sweet notes followed by a lower note and a trill. Call note a *chimp*.

Habitat
Thickets, forest edges, marshes, gardens, and city parks.

Range
Breeds from Alaska to Newfoundland, south to NW. South Carolina, NE. Kansas, New Mexico, and Baja California. Winters from southern half of breeding range to Florida and Mexico.

House Sparrow *Passer domesticus*

The House Sparrow was introduced to North America in 1850, when a few birds were released in New York's Central Park. Since that time the bird, also known as the English Sparrow, has bred prolifically and spread widely, expanding its range to keep in step with the growth of human habitation across the continent. The House Sparrow has proven itself aggressive in the competition with native species for nesting sites, and in some regions this seed-eater has damaged crop yields.

Identification 5½–6¼″. Male streaked above with brown and black, with white wing bar; throat and upper breast black, nape chestnut, and crown gray, with chestnut line through eye. Female streaked brown and black above, dingy gray below, with dull stripe behind eye.

Voice A repeated *chirp*, *cheep*, and various twitters.

Habitat Farmland, cities, towns, and suburban areas.

Range Throughout most of S. Canada and entire United States in cities, towns, suburbs, and agricultural areas.

House Finch *Carpodacus mexicanus*

The House Finch was originally found only in the West; the population established today throughout much of the eastern United States is descended from caged birds set free on Long Island in the 1940s. Sociable birds, House Finches often gather in large winter flocks, but in early spring they form pairs and begin nesting. They are also cheerful singers.

Identification 5–5½″. Male pale brown with darker streaks, and with bright red on forehead, eyebrow, breast, and rump. Female similar but lacks red, and has uniform brown head.

Voice Song a clear, canarylike warble, ending in an ascending *zeeee*. Call note a chirp.

Habitat In the East, in cities and residential areas; in the West, in desert scrub and chaparral.

Range Maine and Michigan south to Georgia; also widespread in the West.

Black-and-white Warbler *Mniotilta varia*

Conspicuous and boldy patterned, the Black-and-white Warbler is often seen circling a tree trunk, heading up to the branches or down toward the ground in search of insects. Because it gleans its food from the bark, this species does not have to wait until the trees leaf out to move northward in spring, and it is consequently one of the first migrant warblers to reach its breeding grounds.

Identification 4½–5½". Black-and-white striped overall; colors in female subdued, and often washed with very pale brown. Throat black in male, white in female.

Voice Song a high, thin, *weesy-weesy-weesy;* call note a *pit* or *zeet.*

Habitat Deciduous forests and damp woodlands; often in coniferous forests in northern part of range.

Range Breeds east of Rockies from N. Alberta to Newfoundland, south to central Texas and North Carolina. Winters from Gulf Coast south to South America.

Pine Siskin *Carduelis pinus*

Unlike the related goldfinches, the Pine Siskin feeds mainly on the seeds of pine trees, which it extracts from beneath the bracts of cones; the bird's slender, conical bill is the perfect instrument for this task. In winter, Pine Siskins make sporadic journeys southward to the United States; their travels often coincide with a shortage of pine seeds in their usual northern haunts.

Identification
4½–5¼″. Brown above and below with darker streaks; some birds are paler overall, others darker; yellow on wing and on deeply notched tail. Females have less yellow.

Voice
A harsh *shick-shick* and a buzzy, ascending *bzzrreeee*. Call note a *sweeeeeet*.

Habitat
Conifer forests; also in alders, aspens, and other deciduous trees near northern conifer forests.

Range
Breeds from S. Alaska to Newfoundland, south through most of forested West and East through Great Lakes region to New England. Winters irregularly farther south.

162

Northern Mockingbird *Mimus polyglottos*

The scientific name of the Mockingbird sums up this species' most remarkable feature. It is indeed a mimic and a polyglot, and its amazing vocalizations are often carried out in the stillness of a humid spring or summer night. It can produce creditable imitations of up to 50 other species of birds; these songs are often given in rapid-fire succession, creating the impression of a landscape alive with bird sounds. It can also imitate tree frogs, barking dogs, crickets, tractors, and sirens.

Identification	11″. Soft gray above, paler gray to whitish below, with white wing bars and long tail. In flight, wings show bright white patches, tail shows white borders.
Voice	Song a rich, infinitely varied medley of harsh and musical notes, interspersed with imitations and repeated phrases.
Habitat	Woodland edges, deserts, cities, suburbs, and farms.
Range	Breeds from N. California to S. Wisconsin and Nova Scotia, south to Mexico and Florida. Northernmost populations move southward in winter.

Downy Woodpecker *Picoides pubescens*

This small woodpecker, common throughout the forests of the East, is fond of visiting backyard suet feeders. Comparatively tame, it occurs widely in residential areas and city parks. The Hairy Woodpecker (*P. villosus*) is found in much the same range, but the shy Hairy is larger, with a longer bill, so there is little chance of confusing the two.

Identification 6–6½″. Black and white above, white below; white cheeks intersected by black eyeline; thin mustache runs from bill to back of neck. Male has small red patch on nape.

Voice Call note a dull *pik*. Also gives a loud, descending rattle. Drums with its bill against bark, producing fast series of percussive noises.

Habitat Forests, woodlands, orchards, residential areas, and city parks.

Range Alaska through most of southern half of Canada, and throughout most of United States; absent from treeless deserts. Some northern birds move south in winter.

166

Northern Flicker *Colaptes auratus*

This species comprises three very distinct color forms that were for a long time considered separate species. The eastern form is still widely known as the Yellow-shafted Flicker. The only woodpeckers that commonly feed on the ground, flickers nest in tree cavities, which they excavate with their powerful bills; like most woodpeckers, they sound trees out for heartrot before they begin their work, as a weakened tree is much easier to excavate.

Identification	11–14″. Brown above with dark spots and barring; buff-white below with black spots and with black crescent on breast; red patch on nape in eastern birds; male has black mustache. White rump and yellow wing linings in flight.
Voice	A loud, repeated *wik-wik-wik* or *flicker-flicker-flicker;* also a loud *kleer.*
Habitat	All habitats with suitable trees for nesting: open country, woodlands, parks, and farm areas.
Range	Breeds throughout North America to northern limit of trees; populations of the Far North fly south in winter.

168

Brown Thrasher *Toxostoma rufum*

Brown Thrashers, members of the same family as the Mockingbird and Gray Catbird, look like thrushes, but they can always be distinguished by their longer tails and streaked rather than spotted underparts. The Brown Thrasher is rather shy and secretive, but it may sing from a conspicuous treetop perch, like its famous relatives.

Identification 11½". Slender. Reddish brown above with 2 white wing bars; buff or pale brown below with heavy streaking on breast and belly. Tail long. Bill long, slightly downcurved.

Voice Song a blend of various musical phrases, each repeated once; somewhat like Mockingbird's. Call note a sharp *smack*.

Habitat Hedgerows, woodland edges, gardens, thickets, and fields with brushy margins.

Range Breeds from Manitoba to Maine, south to E. Texas, Gulf Coast, and S. Florida. Winters in the Southeast and Texas, also along Atlantic Coast to Massachusetts.

170

Wood Thrush *Hylocichla mustelina*

A familiar woodland and garden bird, the Wood Thrush is a cousin of the American Robin; their family resemblance can easily be seen in young Robins, which have a distinct spotted breast. The Wood Thrush is famous for its beautiful, flutelike song, which many people consider to be without rival in nature; Henry David Thoreau wrote of it, "Whenever a man hears it, he is young."

Identification 7½–8½". Brown above, with bright cinnamon-rufous head; white below, with large, rounded black spots. Juveniles similar, but have tawny olive or grayish spots on head and upperparts.

Voice Song a short, flutey *eee-o-lay;* call sharp *pit pit pit.*

Habitat Deciduous woodlands, thickets, and gardens and parks with dense cover.

Range Breeds from Ontario to Nova Scotia, south to Gulf Coast and Mexico. Winters mainly south of United States.

172

House Wren *Troglodytes aedon*

This popular bird sings joyously throughout the breeding season. Besides bird houses and natural cavities in trees, it chooses a great and surprising variety of nesting sites, sometimes taking over old hornet's nests, hats, teapots, mailboxes, old shoes, and even pockets on clothing hung out to dry. House Wrens are fierce competitors for territory, and will pierce the eggs of any other bird that nests in the same area. The House Wren is often seen perched with its short tail cocked up over its back.

Identification 4½–5¼″. Small and plump. Dull brown above, with faint dusky bars on wings; grayish-white below. Tail short, with dark bars and no spots.

Voice Song a rising and falling, bubbling chatter, repeated many times.

Habitat Woodland edges, farms, city parks, and suburbs.

Range Breeds from S. British Columbia, N. Alberta, Ontario, and Maine south to S. California, Arizona, N. Texas, and Georgia. Winters from S. California to Gulf Coast and South Carolina southward.

174

Ruby-throated Hummingbird *Archilochus colubris*

The Rubythroat is the only hummingbird in most of the East. Like other members of its family, it takes nectar from deep within flowers, hovering like a helicopter before colorful blossoms in the spring and summer. In spring, the male performs a spectacular aerial courtship display, swinging like a pendulum through the air while the female watches quietly.

Identification   3¼–3½″. Tiny. Shiny green above, white below; male has iridescent ruby-red throat patch (gorget) that may look blackish from some angles, and black tail. Female lacks gorget, has white tail tips. Bill very long, thin, and needlelike.

Voice   Call a squeaky *chick;* wings hum in flight.

Habitat   Woodlands, gardens, and suburban areas with flowers.

Range   Breeds from central Alberta and through S. Canada, south to E. Texas and Florida. Winters from Mexico south.

176

Guide to Families

Birds are arranged in groups called families and subfamilies. Knowing group characteristics is often helpful in identifying species.

Water Birds

Loons (family Gaviidae) and grebes (family Podicipedidae) are highly aquatic, diving with ease and swimming expertly underwater, but are nearly helpless on land. Loons are larger than grebes; they are dagger-billed birds that nest in the Far North and winter mostly on salt water. Some grebes are widespread inland all year, even on small marshy ponds.

Herons and egrets (family Ardeidae) are long-legged, long-necked, spear-billed birds, usually seen standing in the shallows waiting to snatch fish or other aquatic creatures. Unlike cranes (with which they are often confused), herons and egrets fly with their heads hunched back on their shoulders. Cormorants (family Phalacrocoracidae) are dark, long-necked, long-tailed birds with webbed feet and hooked bills. They pursue their prey by swimming, often below the surface. Pelicans (Pelecanidae) are huge birds with odd bills—long and flat, with an expandable pouch to scoop up fish.

Ducks and Rails

Found worldwide and familiar to everyone, the waterfowl family (Anatidae) includes the swans, geese, and ducks. Waterfowl are sociable and usually migrate in flocks. Geese are often seen feeding on land; the sexes are similar. Male ducks—called drakes—are often brightly patterned, while the females are plain. Dabbling ducks feed with only the head and foreparts submerged, while the diving ducks feed underwater.

The rail family (Rallidae) includes the coots, moorhens, and gallinules—ducklike birds that pump their heads back and forth as they swim, as well as the chickenlike rails, which hide in marshes and are more often heard than seen.

Shorebirds and Gulls Known collectively as shorebirds, the plovers (family Charadriidae) and sandpipers (family Scolopacidae) are mostly brown or gray birds, usually found feeding at the water's edge or in fields. Plovers are short-billed birds with plaintive voices. The more diverse sandpipers (including snipes, godwits, curlews, and others) vary from small to very large, and some have very long bills and legs.

Gulls and terns (family Laridae) are long-winged, mostly gray and white water birds. Most gulls are larger than terns, with blunt-tipped bills and omnivorous habits, usually feeding on the ground or on the water. Terns are

179

graceful aerialists with pointed bills and usually long, forked tails.

Birds of Prey Nature's cleanup crew, the vultures (family Cathartidae) can soar effortlessly for hours, searching for the carrion on which they feed. They have unfeathered heads, heavy bills, and long, broad wings.

The hawks (family Accipitridae) are hunting birds with hooked bills, strong talons, and keen eyesight. This family includes several distinct groups: the broad-winged, soaring buteos; the bird-catching accipiters; the graceful kites; the harriers; and the very large eagles.

The Osprey (subfamily Pandioninae) soars and hovers above the water and dives feetfirst to catch fish in its sharp, curved talons.

Falcons (family Falconidae) are slim hunting birds with pointed wings and long tails, adapted for fast flight.

Typical owls (family Strigidae) and barn-owls (family Tytonidae) hunt by night and roost by day; they share the traits of upright stance, forward-facing eyes, hooked bills, strong talons, and very acute hearing and sight.

Various Landbirds The grouse, quails, pheasants, and turkeys (family Phasianidae) are plump gamebirds, living on the ground, with stout bills and strong legs. Their short, rounded wings carry them on brief bursts of flight to escape danger. The well-known pigeons and doves (family Columbidae) are round-bodied, small-headed, and short-billed; they are often seen walking on the ground.

Wide gaping mouths, tiny feet, camouflaged brown plumage, and nocturnal habits mark the nighthawks, which belong to the nightjar family (Caprimulgidae). In contrast, the hummingbirds (family Trochilidae) are tiny, long-billed, and energetic, with iridescent plumage. Hummingbirds hover before flowers to take nectar and small insects. They are mostly tropical, and only one species reaches eastern North America. The kingfishers (family Alcedinidae) are small-footed, large-headed, dagger-billed birds that dive from a hovering position or from a perch above the water to catch fish.

Clinging to tree trunks with their toes and propping themselves upright with stiff tail feathers, the woodpeckers (family Picidae) use their long, chisel-like bills to seek insects in and below the bark, and to dig nesting holes in dead trees.

Songbirds All of the remaining families are classified as perching birds or "songbirds," although some are not particularly good singers. The flycatchers (family Tyrannidae), including the kingbirds and phoebes, are mostly drab in color but have distinctive voices. They usually sally forth from exposed perches to catch insects. Swallows and martins (family Hirundinidae), while foraging in continuous graceful flight rather than from perches, are usually seen in flocks and often near water.

Named for the odd, waxlike tips on certain wing feathers, the waxwings (family Bombycillidae) are slim, crested birds with soft voices. Living mostly on berries and small fruits, they are highly sociable.

The chickadees and titmice (family Paridae) are small, highly active birds that seek insects in trees from the trunks to the outermost twigs. Most have whistled songs and fussing call notes. Often flocking with them in winter, nuthatches (family Sittidae) are small, short-tailed, chisel-billed birds that clamber up, down, or around tree trunks in search of insects.

The crows, jays, and ravens (family Corvidae) are familiar, large birds with heavy bills, omnivorous habits, and often harsh voices. The wrens (family Troglodytidae) are small, hyperactive birds with thin bills and distinctive voices. They often hold their tails up over their backs or flip them about expressively. The mimic thrushes (family Mimidae), including the mockingbirds, catbirds, and thrashers, are slim birds with long tails and thin, pointed bills. They are often seen feeding on the ground, and are famous for their rich, variable songs. The family Muscicapidae contains several distinct groups. In North America, the old world warblers (subfamily Sylviinae) are represented mainly by the gnatcatchers and kinglets—tiny, insect-eating birds with thin bills. The thrushes (subfamily Turdinae), include the robins, bluebirds, and spotted thrushes; all are fine songsters that feed heavily on fruits and insects. Native to the Old World, the starling family (Sturnidae) includes many colorful members, but our introduced European Starling is rather plain. It shares family traits of sharp-pointed bill and gregarious habits. Vireos (family Vireonidae) are not eye-catching because they wear dull colors and they search methodically among foliage for insects. As if to compensate, all have loud, repetitive songs and scolding call notes.

The family Emberizidae is so large and diverse that it lacks a common name; its many different subfamilies are best considered separately. Small, active, and brightly colored, the wood warblers (subfamily Parulinae) have been called "the butterflies of the bird world." Feeding mainly on insects, they are highly migratory. Warblers provide much of the excitement of spring migration in the East.

Meadowlarks, orioles, bobolinks, grackles, and cowbirds all belong to the diverse blackbird subfamily (Icterinae). They have sharp-pointed bills and omnivorous diets, and usually wear black or warm colors like yellow or orange. Most species occur in flocks outside the nesting season. The subfamily Cardinalinae includes the cardinals, grosbeaks, and buntings. These are small, thick-billed seed-eaters; many species sport bright colors. Similar but usually less colorful are the sparrows and their allies (Emberizinae), including towhees and juncos. Sparrows are most commonly seen feeding on the ground, often in flocks.

In a separate family but similar to birds in the last two groups, the cardueline finches (family Fringillidae) are colorful birds with thick, seed-cracking bills. Most are very sociable and have distinctive flight calls, and many are quite erratic in their migrations. The weavers (family Passeridae) are Old World birds, represented here by the introduced House Sparrow.

On the next page is a list of all the families of birds in this book, together with a page number referring you to the text accounts of the species included.

Water Birds	Loons (36), grebes (34), herons and egrets (54, 56), cormorants (50), pelicans (52).
Ducks and Rails	Ducks and geese (38, 40, 42, 44, 46, 48), rails (32).
Shorebirds and Gulls	Plovers (24), sandpipers (26, 28, 30), gulls and terns (18, 20, 22).
Birds of Prey	Vultures (58), hawks and eagles (60, 64, 66), falcons (68), Osprey (62), typical owls (74, 76), barn-owls (72).
Various Land Birds	Grouse, quails, and pheasants (78, 80), pigeons and doves (82, 84), nightjars (70), hummingbirds (178), kingfishers (132), woodpeckers (166, 168).
Songbirds	Tyrant-flycatchers (134, 138), swallows and martins (124, 126), waxwings (100), chickadees and titmice (140, 144), nuthatches (142), crows and jays (92, 130), wrens (174), mimic-thrushes (122, 164, 170), kinglets (146), thrushes (118, 128, 172), starlings (86), vireos (150), wood warblers (104, 108, 110, 116, 160), blackbirds and their allies (88, 90, 94, 102, 114), cardinals, grosbeaks, and buntings (96, 98), sparrows, towhees, and juncos (120, 136, 148, 152, 154), finches (106, 112, 158, 162), weavers (156).

Glossary

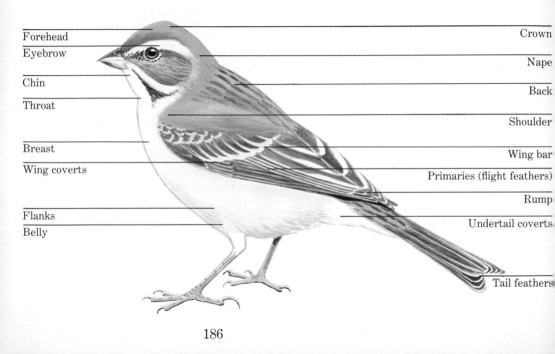

Forehead

Eyebrow

Chin

Throat

Breast

Wing coverts

Flanks

Belly

Crown

Nape

Back

Shoulder

Wing bar

Primaries (flight feathers)

Rump

Undertail coverts

Tail feathers

Color phase
One of two or more distinct color types within species, occurring independently of age, sex, or season.

Coverts
The small feathers covering bases of usually larger feathers, providing a smooth, aerodynamic surface.

Crown
The upper surface of the head, between the eyebrows.

Eye-ring
A fleshy or feathered ring around the eye.

Eye stripe
A stripe running horizontally from base of bill through eye.

Mandible
One of the two parts, upper and lower, of a bird's bill.

Mask
An area of contrasting color on front of face and around eyes.

Nape
The back of the head, including the hindneck.

Rump
The lower back, just above the tail.

Speculum
A distinctively colored area on the rear edge of the wing of many ducks.

Underparts
The lower surface of the body, including the chin, throat, breast, belly, sides and flanks, and undertail coverts.

Wing bar
A bar of contrasting color on the upper wing coverts.

Wing lining
A collective term for the coverts of the underwing.

Wing stripe
A lengthwise strip on the upper surface of the extended wing.

187

Index

Photographers

Ron Austing (65, 103, 141, 169, 177)
Robert H. Armstrong (105)
S.R. Cannings (163)
N.R. Christensen (33)
Betty Darling Cottrille (111, 117)
Adrian J. Dignan (147)
Larry Ditto (25, 35, 95)
Harry Engels (23)
Jon Farrar (43)
Tim Fitzharris (27)
Joseph A. Grzybowski (151)
James M. Greaves (155)
Francois Gohier (59)
William D. Griffin (47, 63)
Isidor Jeklin (75, 79, 83, 129, 133, 167, 171)
G.C. Kelley (49)
Wayne Lankinen (67, 101, 113, 131, 135, 143, 145)
Frans Lanting (61)
Thomas W. Martin (87, 89, 109, 121, 127, 139)
Joe McDonald (77, 97)
C. Allan Morgan (45)
Jerry R. Oldenettel (51)
James F. Parnell (29)
Jan Erik Pierson (19)

John C. Pitcher (37)
Rod Planck (115, 119)
Leonard Lee Rue III (81, 125, 159)
C.W. Schwartz (73, 93)
John Shaw (57, 153)
Perry D. Slocum (55, 175)

Tom Stack & Assoc.
John Shaw (71)

Alvin E. Staffan (137)
Frank Todd (39, 41, 53)
John Trott (107, 149, 161)
Robert Villani (91, 157)
Richard E. Webster (31, 85)
Wardene Weisser (69)
Larry West (3, 17, 99)
Jack Wilburn (21, 165)
Leonard Zorn (123, 173)

Cover Photograph
Cedar Waxwing
by Wayne Lankinen

Illustrators
Range Maps by Paul Singer
Drawing p. 186 by Lars Svensson

Prepared and produced by Chanticleer Press, Inc.
Publisher: Andrew Stewart
Managing Editor: Edie Locke
Senior Editor: Amy K. Hughes
Editorial Assistant: Kristina Lucenko
Art Director: Drew Stevens
Production: Alicia Mills
Photo Editor: Giema Tsakuginow
Photo Assistant: C. Tiffany Lee

Founding Publisher:
Paul Steiner

Staff for this book:
Editor-in-Chief: Gudrun Buettner
Executive Editor: Susan Costello
Managing Editor: Jane Opper
Senior Editor: Ann Whitman
Natural Science Editor: John Farrand, Jr.
Associate Editor: David Allen
Assistant Editor: Leslie Marchal
Production: Helga Lose, Gina Stead
Art Director: Carol Nehring
Art Associate: Ayn Svoboda
Picture Library: Edward Douglas

Original series design by Massimo Vignelli

All editorial inquiries should be addressed to:
Chanticleer Press
568 Broadway, Suite #1005A
New York, NY 10012

To purchase this book, or other National Audubon Society illustrated Nature Books, please contact:
Alfred A. Knopf, Inc.
201 East 50th Street
New York, NY 10022
(800) 733-3000

NATIONAL AUDUBON SOCIETY

The mission of the NATIONAL AUDUBON SOCIETY *is to conserve and restore natural ecosystems, focusing on birds and other wildlife, for the benefit of humanity and the earth's biological diversity.*

With more than 560,000 members and an extensive chapter network, our staff of scientists, educators, lobbyists, lawyers, and policy analysts works to save threatened ecosystems and restore the natural balance of life on our planet. Through our sanctuary system we manage 150,000 acres of critical habitat. *Audubon* magazine, sent to all members, carries outstanding articles and color photography on wildlife, nature, and the environment. We also publish *Field Notes*, a journal reporting bird sightings, and *Audubon Adventures*, a bimonthly children's newsletter reaching 600,000 students.

NATIONAL AUDUBON SOCIETY produces television documentaries and sponsors books, electronic programs, and nature travel to exotic places.

For membership information:

NATIONAL AUDUBON SOCIETY
700 Broadway
New York, NY 10003-9562
212-979-3000